THE EYE
A LIGHT RECEIVER

Author
Wilburn L. Sooter

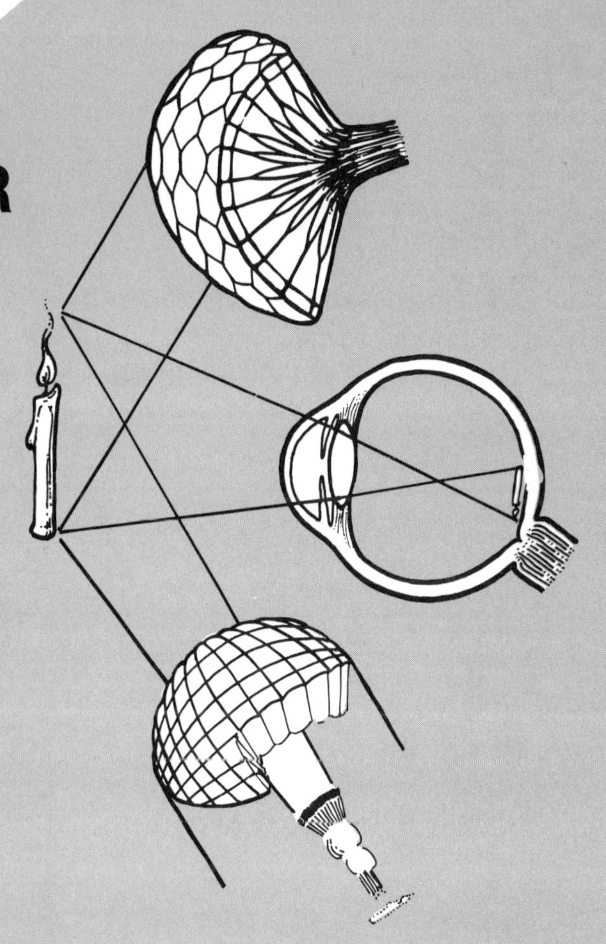

Project Director
Richard B. Bliss, Ed.D.

Illustrator
Linda Vance

CLP PUBLISHERS
SAN DIEGO, CALIFORNIA

THE EYE: A LIGHT RECEIVER

Published and distributed by
CLP PUBLISHERS, P. O. Box 15666, San Diego, California 92115

Produced and developed as part of a writing project sponsored by the **INSTITUTE FOR CREATION RESEARCH**

Richard B. Bliss, Ed.D., Project Director

CONSULTANTS

Theodore Fischbacher, Education, Ph.D.
Jean Sloat Morton, Biology, Ph.D.
Gary E. Parker, Biology, Ed.D.
Hazel May Rue, Education, M.S.
Harold S. Slusher, Physics, M.S., D.Sc.

CONTRIBUTING TEAM MEMBERS

ILLUSTRATORS

Shirlene Barrett, Jonathan Chong, Richard Holt, Doug Jennings, Karen Myers, Marvin Ross,
Barbara Sauer, Linda Vance, Frankie Winn

PROJECT WRITING STAFF

Deborah Bainer—Malaysia
Anne Beams—Germany
Gary G. Eastman—California
Elizabeth Ernst—Oregon
Kenneth F. Ernst, Jr.—Oregon
Olive Fischbacher—California
Norman Fox—Oregon
Virginia Gray Hastings—Illinois
Marilyn F. Hallman—Texas
Alberta Hanson—California

Richard Holt—Iowa
Melody J. McIntyre—Pennsylvania
Fred Pauling—Virginia
Hazel May Rue—Oregon
Barbara Sauer—Illinois
Wilburn Sooter—Washington
Ivan Stonehocker—Canada
Harold G. Watkins—California
Susan E. Watkins—California
Fred Willson—California

THE EYE: A LIGHT RECEIVER

Copyright © 1981
Institute for Creation Research
El Cajon, California 92021

Library of Congress Catalog Card Number 81-68313

ISBN 0-89051-076-8

Cataloging in Publication Data

The eye, a light receiver / Wilburn L. Sooter ; Linda Vance, ill.

1. Eye—Study and teaching (Elementary)
2. Vision—Study and teaching (Elementary)
I. Title.

612.84

ISBN 0-89051-076-8   81-68313

Printed in the United States of America

CONTENTS

PRONUNCIATION KEY

The following pronunciation key is based on the Thorndike-Barnhart School Dictionary. These markings are used in your margin glossary to help you pronounce important words. ·

hat, āge, fär

let, ēqual, térm

it, īce

hot, ōpen, ôrder, oil, out

cup, put, rüle

ch, child

ng, long

sh, she

th, thin

ℸh, then

zh, measure

uh represents *a* in about, *e* in taken, *i* in pencil,

o in lemon, *u* in circus

INTRODUCTION TO THE STUDENT

Mr. Light Ray wants to take you on a journey through the eye.

Read carefully, look at the drawings, and think about each part as you read.

Did the eye, as we see it today, develop over a long period of time by chance? Or was it designed by a master designer? Some scientists believe that the study of the eye provides a means of seeing a creator. Other scientists believe the study of the eye favors the process of evolution. We hope that you will be stimulated to think about the data and decide which **model** seems more likely to be correct.

model (mod´ l) viewpoint

A study made in 1978 showed that students appear to learn more about facts of origins when they study from a two-model approach. The study seemed to show that the students' ability to think through ideas was greatly strengthened. In fact, these students seemed to be more highly motivated than those who were taught only one model.

Have you watched a hawk soar high in the sky? Do you know what he is doing? He is probably looking for his food. He has a special kind of eye that can see small mice and other rodents hundreds of feet below. When he sees his dinner, he drops to a low altitude, folds his wings, and dives silently and swiftly. The hawk has one kind of special eye. The bee has another special kind of eye to help him do his work. He sees different colors that help him find his food. Cows and horses see only black and white. A squid can see in very dim light with his special kind of eye. All eyes must have light to see. Light must reflect off something and be picked up by the eye.

A LIGHT RECEIVER

The eye is a light **receiving** instrument. Light and light receivers are a fascinating and important part of our universe. Without a light receiver in our body, this world would be a dark place in which to live.

receiving (ri sēv´ ing) taking in light

There are many amazing things about light and light receivers that are hard for the average person to explain. Therefore, we are going to introduce you to a special light beam named Ray. Mr. Light Ray has been around a long time and has traveled into many light receiving instruments. He knows them inside and out. We are going to ask him to tell us about the eye.

Here's Ray now!

I am very glad you want to study the eye. The eye is a light receiver. Its purpose is to make use of me. I have been in your eye many times. I always enjoy talking about my travels in the eye.

I really don't have a light bulb head. I look like a series of dots in the form of a wave.

To show you through the eye, I am going to look like a simple line.

There are three types of eyes that I will show you. They are the human or **simple eye**, the **compound eye**, and the **superpositioned compound eye.**

First, I will take you through the human eye. I will show you how I travel through it and explain its parts. We will also think about how it developed, or was made into such a fine instrument.

Our journey will take us through the **cornea**, then through the **pupil** and **iris**, and on through the **aqueous humor** to the **lens**. From there we go through the **vitreous humor** to the **retina**. I can go no further than this. In the retina I cause chemical impulses that carry the image down the **optic nerve** to the brain.

I will explain each part as I come to it.

simple eye (sim´ puhl ī) contains one lens that concentrates a light image upon a retina (on back of eye ball)

cornea (kôr´ nē uh) outside surface of the eye we see

pupil (pyü´ puhl) hole in the iris of the eye

iris (ī´ris) colored part of the eye

aqueous humor (ak´ wē uhs hyü´ muhr) fluid between the cornea and the lens of the eye

lens (lenz) a clear, glass-like material that forms an image by focusing rays of light

vitreous humor (vi´ trē ühs hyü´ muhr) clear, jelly-like material in the eyeball to give it the required shape

retina (ret´ n uh) inner lining of the back of the eyeball

optic nerve (op´ tik nėrv) a bundle of nerves that go from the back of the eye to the seeing center of the brain

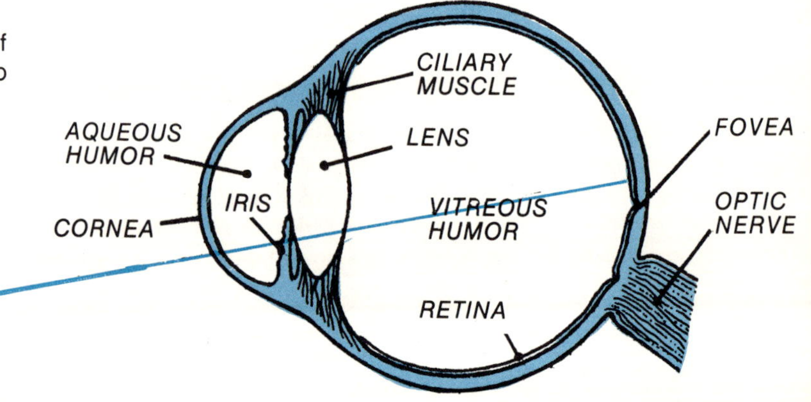

4

I am now entering the **cornea**, the outside surface of the eye. It is a **transparent** membrane or skin. Notice that it is curved. This helps direct me through it correctly. Sometimes it is called the white of the eye. Contact lenses are often put on the cornea. The cornea is touched by the eyelid.

Washing and **lubricating** is needed on the delicate surface of the cornea. The **tear gland** secretes salty fluid that does this. Where do you think the tear gland is located? It is found above the upper surface of the eyeball. When you are in pain or afraid, the tear gland produces an excess amount of fluid. The tubes that take off the normal flow into the nose will then overflow and drop off in tears and your friends may ask you why you are crying.[1]

Have you ever seen your dog crying? Your pet does have tear glands, but they do not overflow. People are the only ones who can cry.

transparent (tran sper′uhnt) material you can see through

lubricate (lü′bruh kāt) to make smooth or slippery

gland (gland) group of cells that do a special work

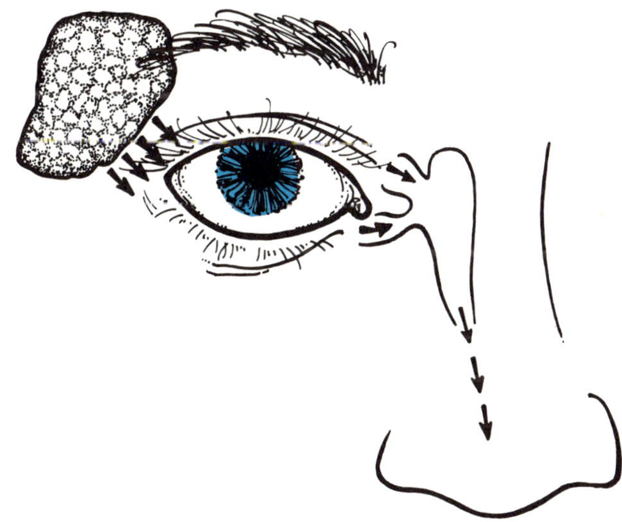

From the cornea I enter a black circle called the **pupil**. It is made to open and close automatically and is triggered by the amount of my rays coming into it. If a great many light rays come at the pupil it will nearly close. Take away the light and it opens wide.

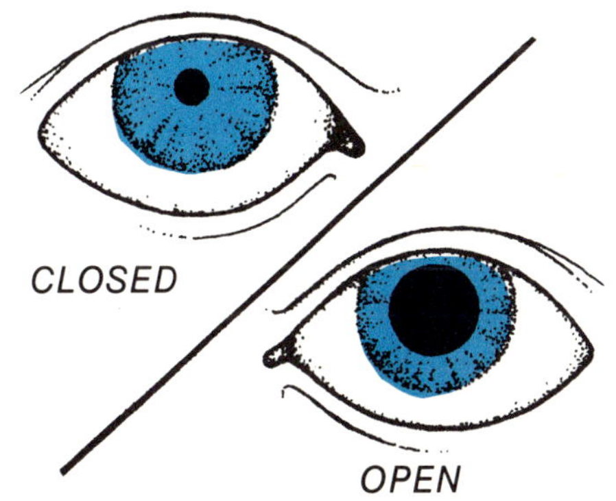

Not all living beings have the same kind of pupils. People have round pupils. They are able to see equally in all directions.

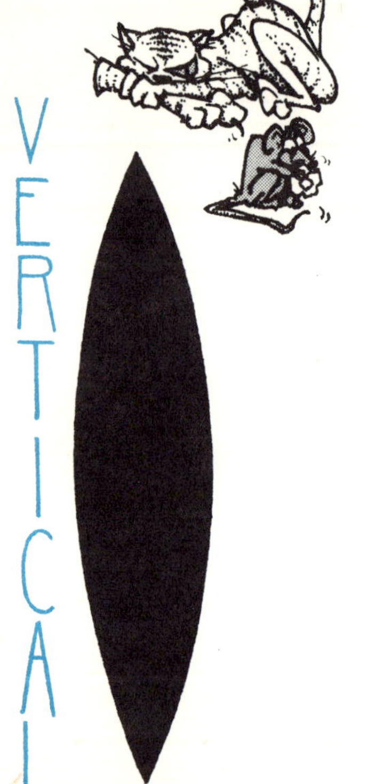

The pupil of a cat's eye is a vertical slit. He is able to see best up and down.

The horse has a horizontal slit pupil, so he can see best from side to side.[2]

Around the pupil is the **iris**. It allows the pupil to become large or small. Iris means rainbow in Latin and it may be any color. If your iris is blue, you have blue eyes. If it is brown, you have brown eyes.

Isn't the structure of the eye exciting? It not only lets us see, but has color, opens and closes to my light rays, washes and lubricates itself, and has various types of pupils to suit the needs of animals and man. It is a fantastic instrument!

Notice that the cornea is round. The shape of the eye is important to its function. Between the cornea and the lens is a fluid called the **aqueous humor**. It is 98% water and is much like blood plasma. It feeds the cornea its food and makes the cornea round and firm. A round shape is necessary for this area.

For you to see clearly, the reflected light from an object must be focused clearly as an image on the back of your eye. An instrument used to **focus** light rays is called a lens. The lens of the eye is a **crystalline structure**. It is about the size of a small lima bean. The **ciliary muscles** surround the lens to hold the lens in place and make it thick or thin.

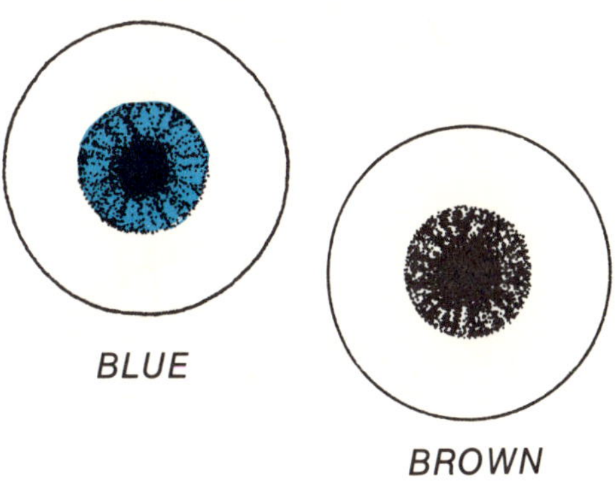

image (im´ ij) just like something else

focus (fō kuhs) (a) point at which rays of light come together, or separate to make an image; (b) to look at something

crystalline (kris´ ll uhn) colorless, transparent material

ciliary muscle (sil´ ē er´ē mus´ uhl) muscles controlling the lens

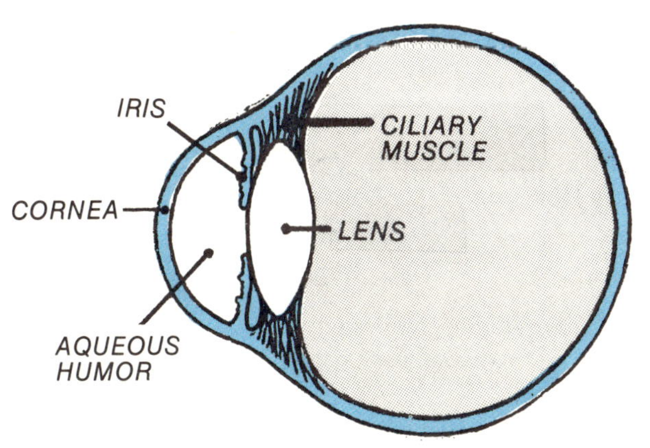

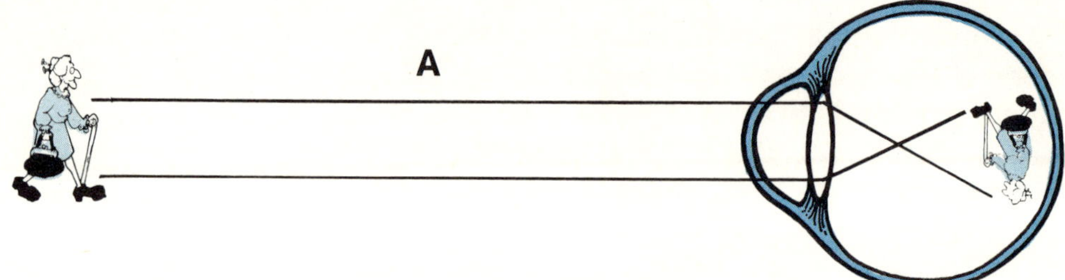

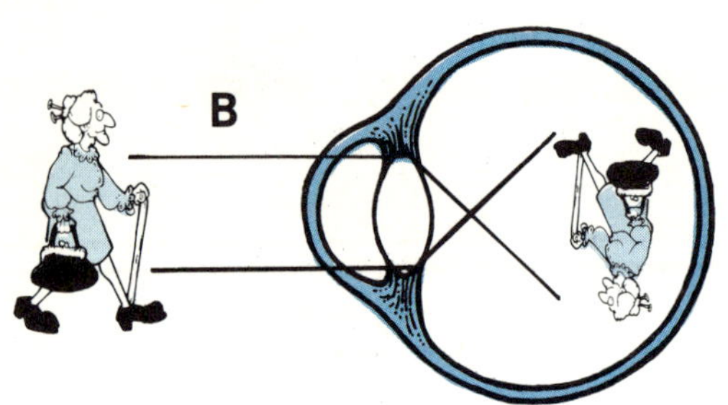

Any light ray knows that shape is important to the eye. Shape is also very important to the work of the lens. When the ciliary muscles make the lens thin (see drawing A), we light beams coming in from far away are focused clearly on the back of the inside of the eye. When the lens is thick (see drawing B), the light beams close together, but are still focused clearly on the back of the inside of the eye.

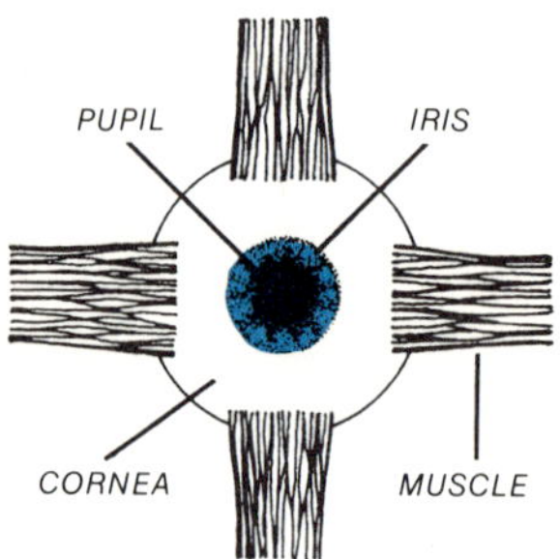

Another muscle structure in the eye controls the movement of the eye up, down, and from side to side. You can see the attached muscles in the drawing.

Again, shape is important as I travel to the inner eye. Between the lens and the back wall of the inside of the eye is a material similar to the aqueous humor. It is called the vitreous humor. It does three things:

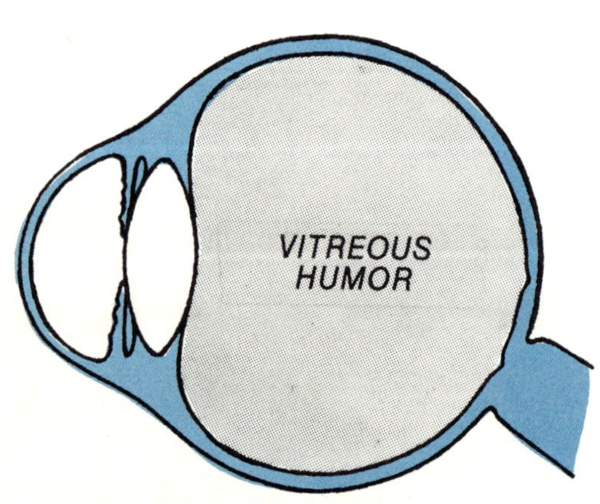

1. It gives the eyeball its round shape.
2. It allows me to pass to the back inside wall of the eye without being changed.
3. It helps feed the lens and inner eye its food.

Just as a movie projector shows an image on a screen, the lens focuses an image on the back side of the inside of the eyeball. The image is always upside down.

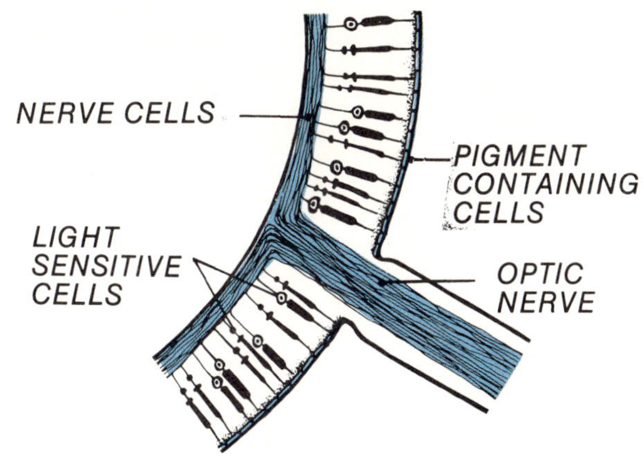

pigment (pig´ muhnt) color

The back wall of the eye has a special name. It is called the retina. The retina has three important layers:

1. Nerve cells which allow us light rays to pass through them.

2. Light-sensitive cells that capture us.

3. **Pigment**-containing cells with dark layers to prevent us from scattering around in the inner eye.

REVIEW

Light rays cannot go any further into the eye than the retina.

This is a good place to stop and review my path through the eye. Let's also do some thinking about how the eye was made.

You see something because light reflects from it. The first part of the eye that I hit is the cornea. Then I pass through the iris and pupil, and on through the aqueous humor to the lens. Next I pass through the vitreous humor to the retina. The image is then placed on the retina and taken to the brain by chemical impulses.

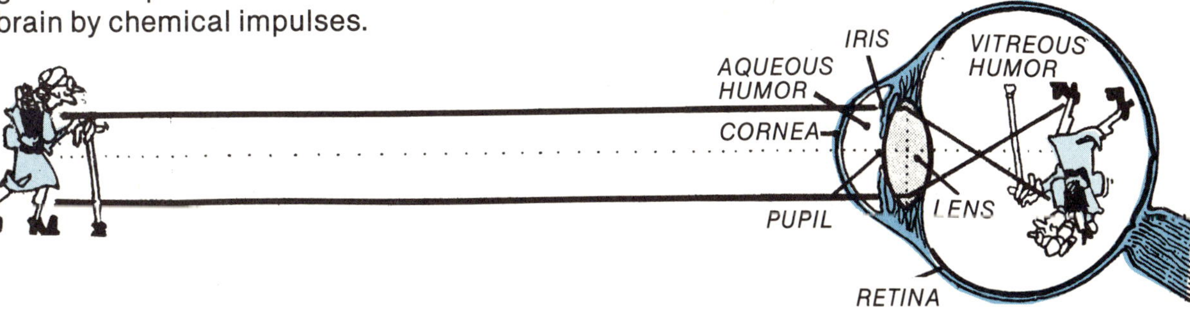

The parts of the eye are not just hard metal pieces made by some man to do some job. They are living cells that need food and care. They can repair themselves. Man has not yet been able to make a machine that repairs itself. The parts of the eye work closely with each of the other parts. This lets you see where you are going and enjoy other people and things.

TWO WAYS TO THINK

Scientists observe and think about what they see. One scientist made statements something like these as he studied the eye. Nature has produced incredible shapes in the eye. It was no accident. He further went on to say, "The lenses look like they were designed by a physicist."[3]

We could think of these statements in two different ways. Did nature produce the eye by evolution or did a designer with a mind make the eye? "Who made the eye?" or "What made the eye?" are simple ways of asking the question. Evolutionist and Creationist scientists both see the same facts, but they look at these facts in different ways.

scientist (sī´uhn tist) a person learned in science

creation (krē ā´shuhn) the idea that all life forms are varieties of kinds created by a master designer

evolution (ev´uh lü´shuhn) the idea that simple life forms can change to complex ones over long periods of time

CREATION

A **Creationist** would say that the eye was planned. Each part was designed for its special purpose. The parts were put together and hooked up in the body with nerves and blood vessels to receive the image and interpret the image in the brain.

EVOLUTION

The **Evolutionist** would say the eye may have developed by gradual change. Each step probably gave the eye a slightly better vision. It may have begun only as an area which could tell light from dark. The blurred image came first, then a primitive lens. Later improvements led to clearer vision. The eye adapts to the needs of the animal and by the method of trial and error the poor parts did not carry over and the better parts did. Slight changes to improve the eye continued. The present eye is the result of this gradual change. This happened over a period of millions of years.[4]

What makes sense to you? Did a Creator make the eye? Or did time and chance develop it? Think about these two ideas that scientists are considering as they look at the eye. Science forces you to wonder how it all began.

Let us finish looking at the eye now that we have thought about it from two viewpoints.

A light ray cannot go any further into the eye than putting the image on the retina. So I will tell you what happens from the retina to the brain.

nerve cells (nėrv selz) the cells in the body that transmit messages to the brain

connectors (kuh nek′ tėrz) nerves that go from one part of the body to another

receptor (ri sep′ tuhr) a cell that takes in light

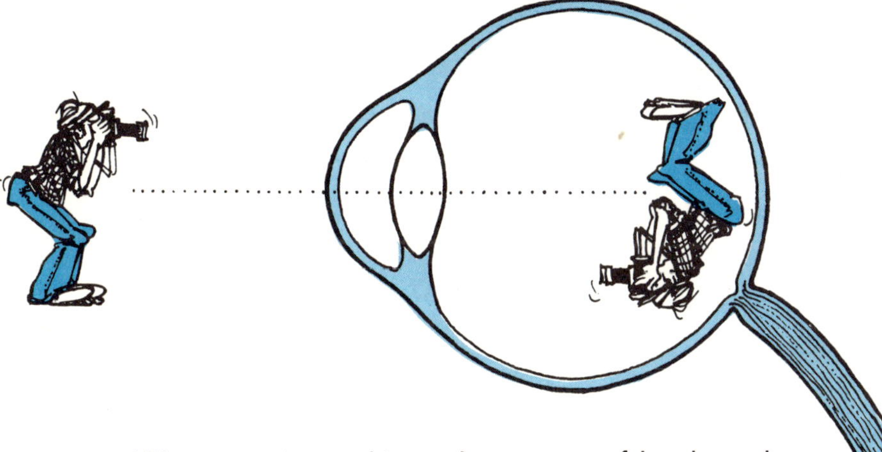

When we stopped to review, my ray friends and I were just making an image on the retina. In order for you to see, the image must be sent to the brain. The retina contains **nerve cells** that are the **connectors** between the brain and the retina. **Receptors** pick up the light from the image on the retina. They change the light to chemical energy and carry it to the brain. There are about 150 million receptor cells in the retina. These millions of cells work together to enable you to see.

rods (rodz) nerve cells in the retina that pick up dim light

cones (kōnz) nerve cells in the retina that pick up color

There are two kinds of receptors called **rods** and **cones**. Rods are long and thin. Cones are wider and slightly longer. They look something like this as they are in the retina.

Rods and cones have a special pigment that decomposes as soon as light falls on it, then it reforms at once with the aid of some body chemicals and vitamin A.[5]

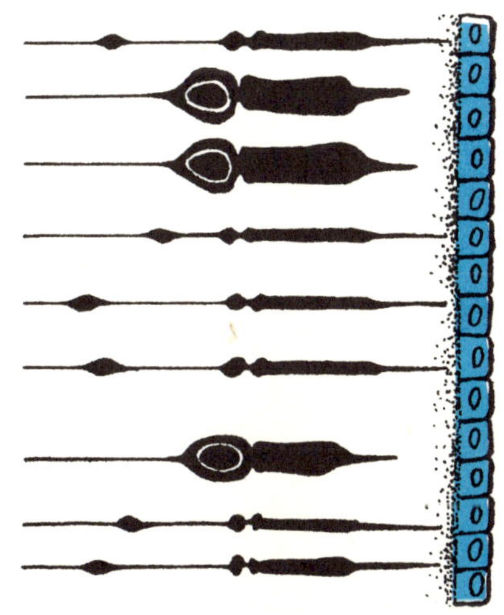

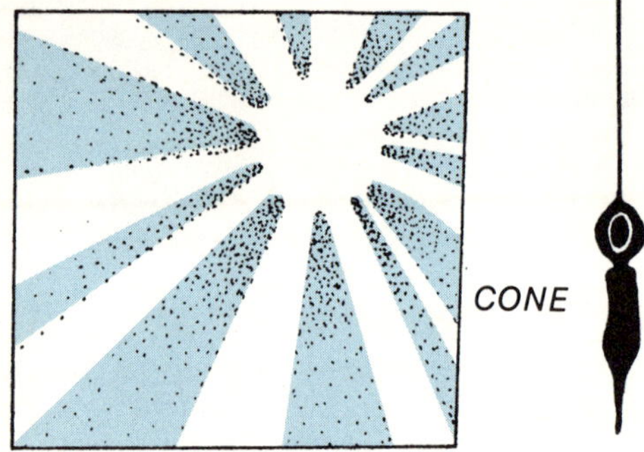

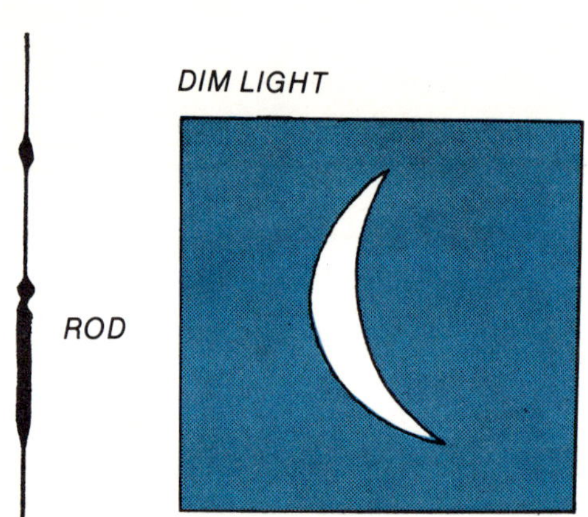

Rods and cones have jobs to do that are different from each other. Rods are thinner than cones and allow us to see in dim light. Cones pick up color, shapes, and details in bright light.

The rods and cones are set in a black covering of the retina. This black covering helps prevent the spread of the image to other places in the inside of the eyeball.

The old woman in the drawing below is the image on the retina. Receptor cells in the retina pick up her image and make it into chemical energy. The chemical energy sets up a series of plus (+) and (−) minus electric charges in the nerve cells called impulses. The nerve cell is connected to the optic nerve, where the impulse is taken to the brain. The optic nerve is a bundle of individual nerves leading from the back of the eye to the brain. All the receptor cells in the retina are connected to the optic nerve.[6]

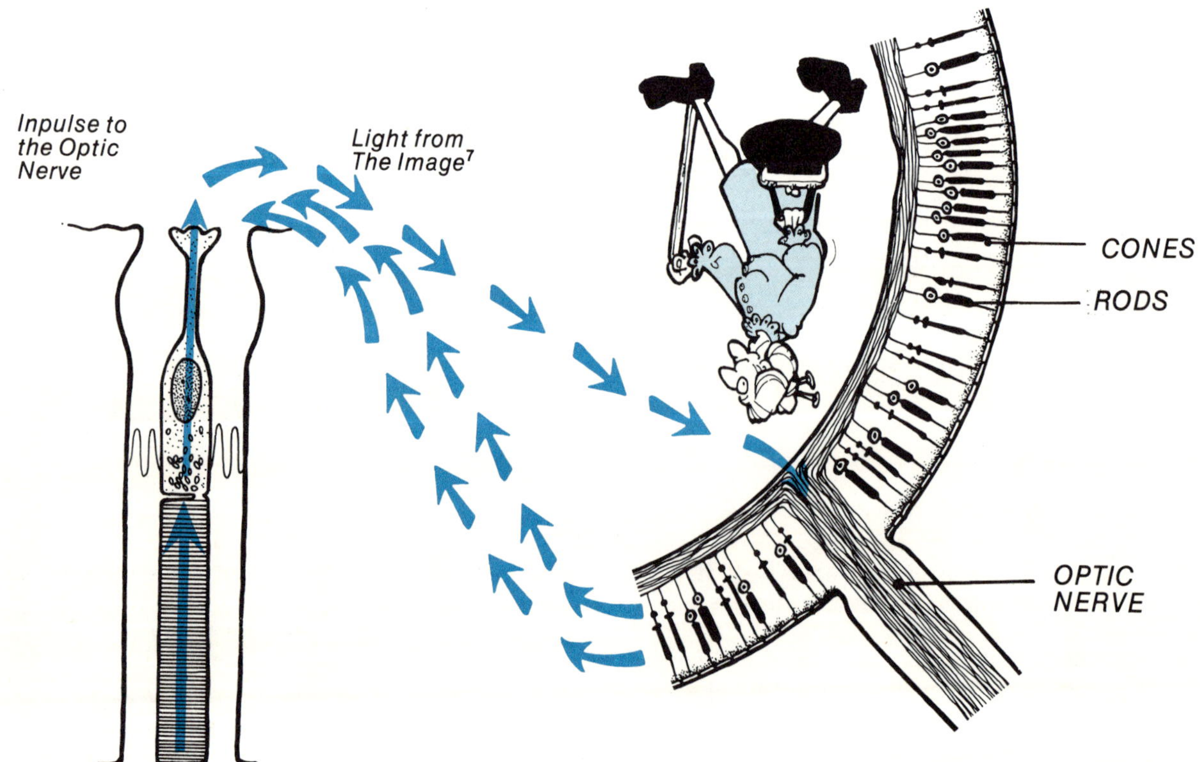

THINKING ABOUT SEEING

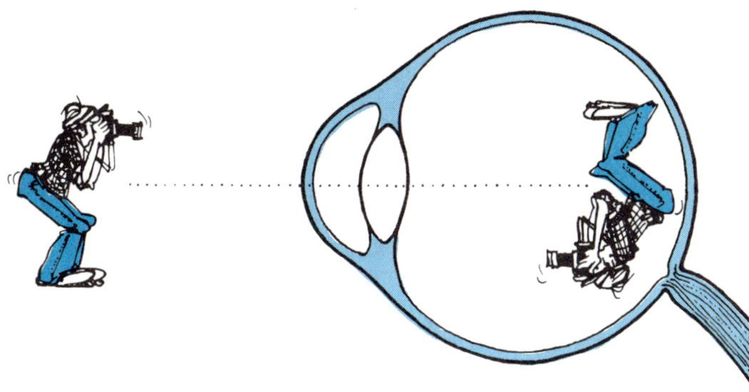

On this page you will be challenged to think! Read it through and decide what you think is correct.

What part of the eye does the seeing? Does your eye see or does your brain see? Close your eyes and try to imagine a ball or your jump rope. Can you remember how they look? Does your brain recall how they look or does your eye recall the image? Where does seeing really take place? The brain **decodes** the message and you see the object that **reflected** the light to the retina.

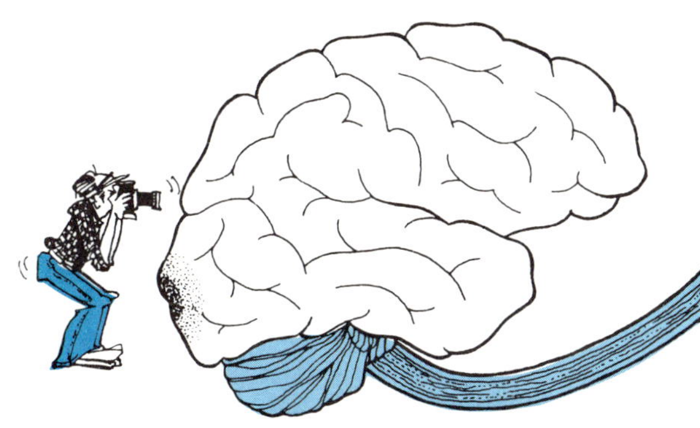

decode (dē kōd́) to give understanding

reflected (ri flek´ ted) bounced back

Seeing really takes place in the brain with the help of the eye and Ray, the light source.

Did you know the eye has a special area where cells are concentrated for reading and fine work? This special place in the retina is called the **fovea** which contains cones only. They are closer together in the fovea, giving clearer vision.[8] My light focuses the image for reading and doing very detailed work on the fovea so you can see small things more easily.

The large circle on the drawing below shows the optic nerve coming out from the retina. No nerve cells or receptors are located in this disk. Do you think you can see if an image falls on this spot? Did you say, "No"? Then you are correct. Everyone has a "blind spot" in his vision if an image falls on this area of the retina. Look in the appendix and you will find a way to test your range of vision and find your blind spot.

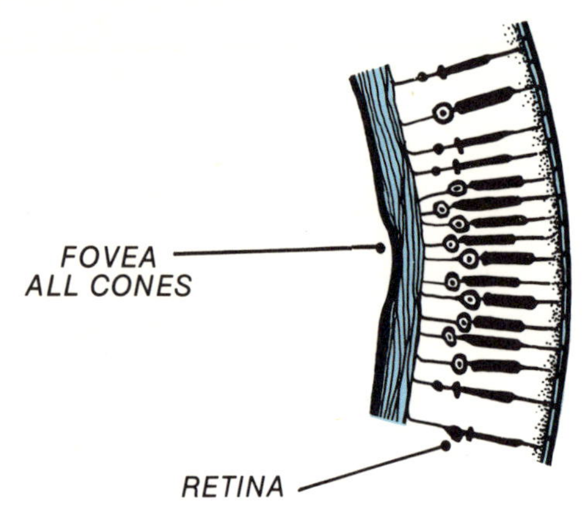

fovea (fō′vē uh) area in retina where there are more cones and where vision is the clearest

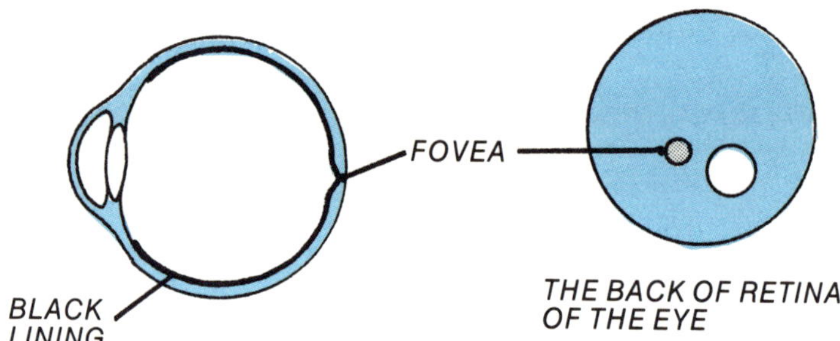

Why do you think we have two eyes instead of one? I hope you guessed it. We need two views of our world. Why do we need two views of the world? In the box below are some reasons we need two views through our two eyes.

Let's think of some reasons two eyes are better than one. First, we can see depth and shape. Second, we have more accurate vision. Third, we can see from a wider angle. Can you think of some other reasons?

The most important reason we have two eyes, however, is so we can see depth and shape. Sometimes this is called **binocular** vision. We see two slightly different pictures. One with each eye. They are put together in the brain as one picture with depth and shape. A camera sees only one picture because it has only one eye. Therefore, a snapshot is flat.[9]

Our eyes are movable, and we can move our heads. With movement and the memory of the brain for seeing, we can sometimes visualize depth with one eye.

Would you like to see two single images put together into one? Try this. Hold this book in front of your nose. Close one eye. You see one side of the book. Now close the other eye, instead. Now you see the other side of the book. With both eyes open, depth is seen. Binocular vision puts two images together and you see in depth. The same idea is used in a 3-D picture viewer, such as a "View Master." Each eye sees a different angle and it gives depth perception.

binocular (buh nok´ yuh luhr) two pictures blend into one to see depth and shape

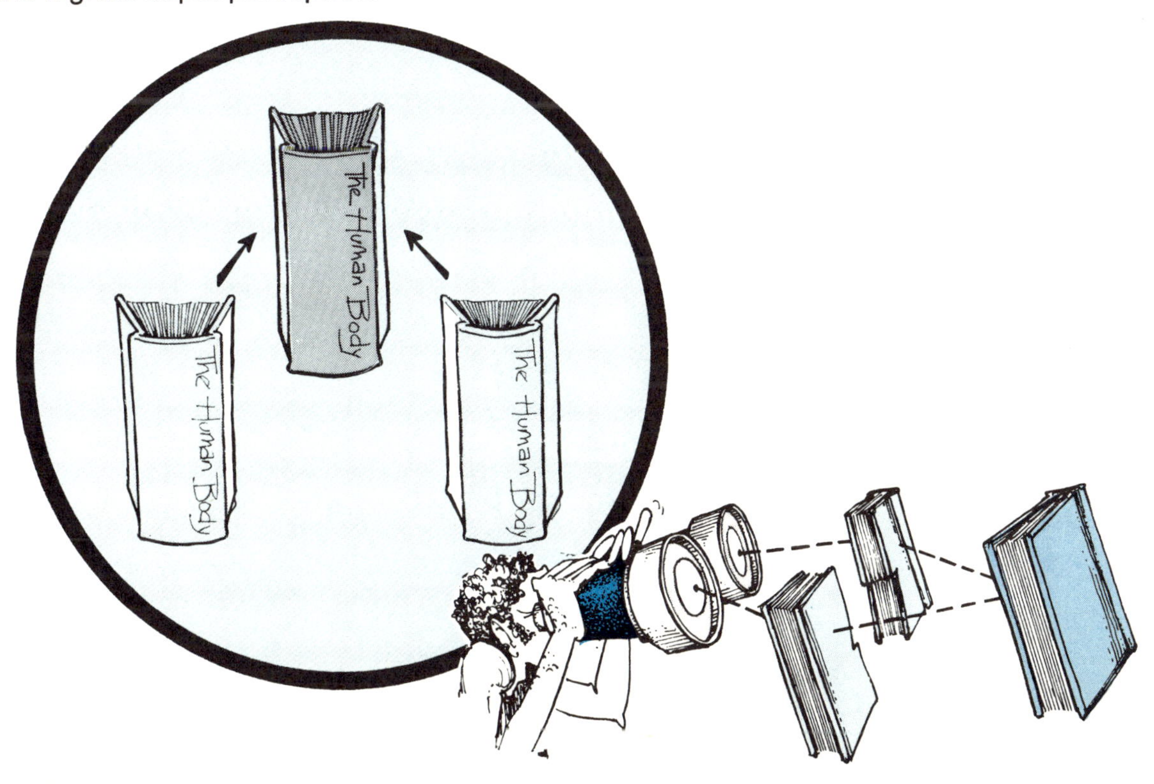

OTHER TYPES OF EYES

Scientists call the human eye the simple eye. Of course, it is not simple in structure, but very complicated. It is called "simple" because we have only two of them.

Did you know that insects have hundreds of eyes? Some insects' eyes have six-sided lenses and others have square lenses. Besides the simple eye, there are two other kinds of eyes, the **compound eye** and the **superpositioned compound eye**. The compound eye is the eye of daytime insects. The superpositioned compound eye is the eye of some nighttime insects.

compound eye (kom′ pound ī) eye with many facets and no retina

superpositioned compound eye (sü′ puhr ˌpuh zish′ uhnd kom′ pound ī) eye with many facets and a retina

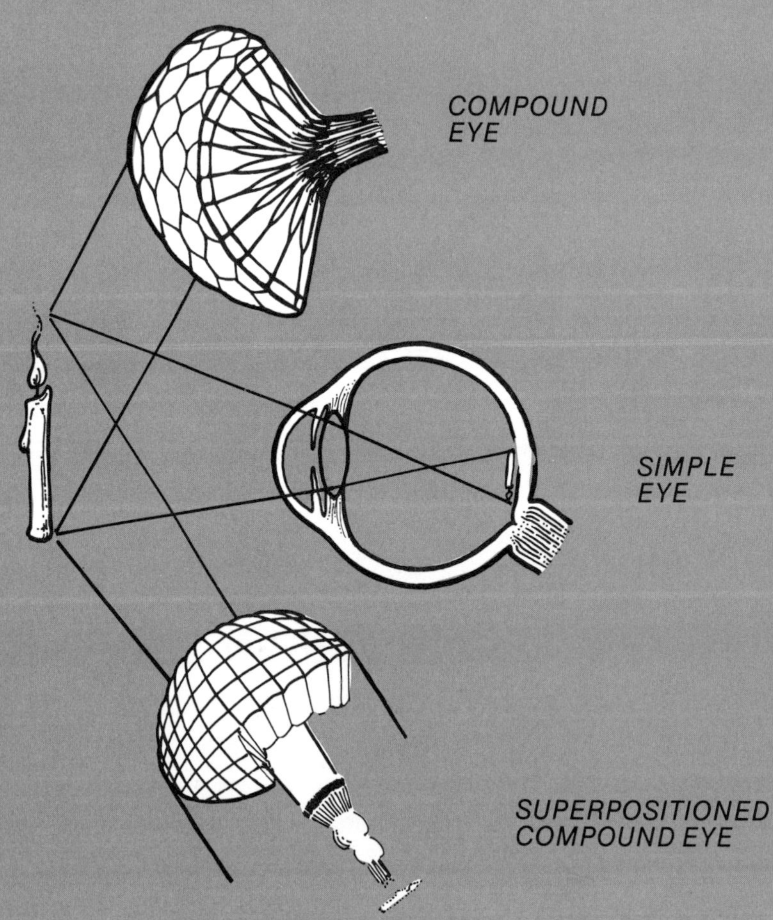

THE COMPOUND EYE

Mr. Light Ray

Look at the lens. Each has six sides; that is, it is hexagonal in shape. Each lens picks up its portion of the view of the object. I enter the lens and cause the nerve cells to relay impulses to the optic nerve and on to the brain. The brain puts all the little images together for a complete picture.

The insect eye is larger than its brain. It is mounted on the head so it can see in all directions. It is able to see more clearly in the front because it has larger hexagonal lenses pointing forward,[10] picking up more of my light.

SUPERPOSITIONED COMPOUND EYE

Few people have heard of the second kind of insect eye called the superpositioned compound eye. It has special parts that can pick up very dim light. Very small rays, like me, enter the lens faintly and cause impulses to move down the nerve cells. The superpositioned eye has square lenses or facets. It has a retina like your eye, but it is much smaller.

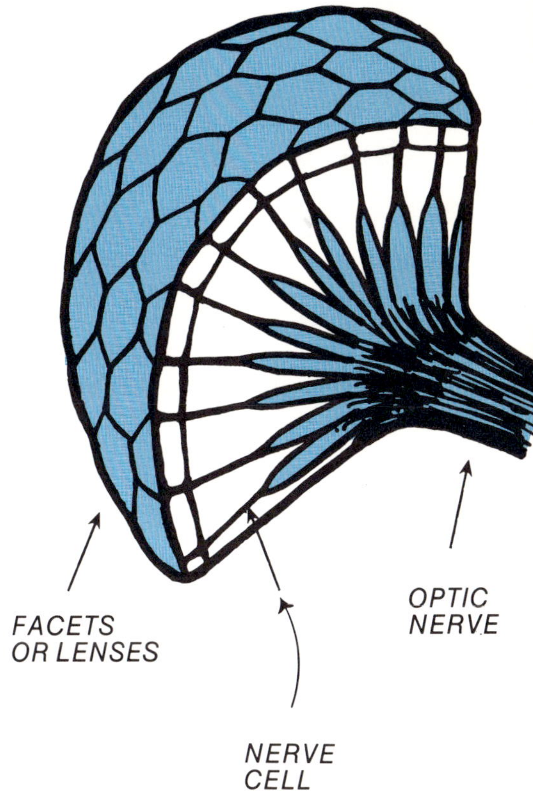

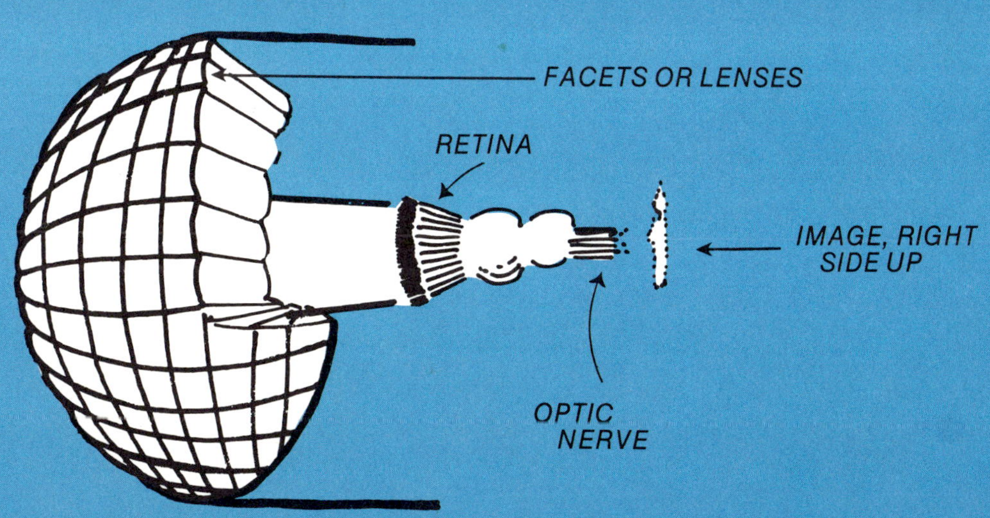

There are two ways the superpositioned compound eye has of collecting more light to make an image on its retina:

1. All the facets collect us light rays and focus us all together on the small retina.
2. The length of the square facet enables it to collect more light.[11]

It is able to take a very little light from a large area and bring it together in a small area. The animal is then able to see in the dark.

Another special thing is found in the superpositioned eye. The image is right side up on the retina instead of upside down, as in the simple eye. Let us compare the two kinds of eyes by listing some likenesses and differences:

ALIKE

They are alike by

1. Having lenses on the outside.
2. Having receptor cells.
3. Having a nerve system.
4. Having a brain to interpret images.
5. Being an equally complicated optic system.

DIFFERENT

The differences are:

1. One has square facets.
2. One has hexagonal facets.
3. The superpositioned eye has a small retina.
4. The compound eye has no retina.
5. The superpositioned eye has a right-side-up image on the retina.
6. The superpositioned eye has the ability to collect light beams from a large area and focus them on a small retina.

REVIEW

Hasn't it been exciting to trace my travels through the eye? Millions of cells make up the eye. All the cells are set in perfect order. Each cell does its job well. I want to take you on a quick review tour and then do some more thinking about the eye and its beginning.

I enter the eye through the cornea, the outside surface of the eye. Then I go on through this colored iris by way of the black pupil. Next I pass through the liquid aqueous humor to the lens. The lens focuses me and other light rays on the retina. When focusing, the lens is made thin by the ciliary muscles. My focused ray passes through the vitreous humor to the retina where the image is projected upside down. I can go no further as a light ray than to the retina. The receptor cells of the retina take us light rays and make us into chemical impulses. The chemical impulses move down the nerve cells to the optic nerve and on to the brain. The brain interprets the impulses, and the person sees the object reflecting the light.

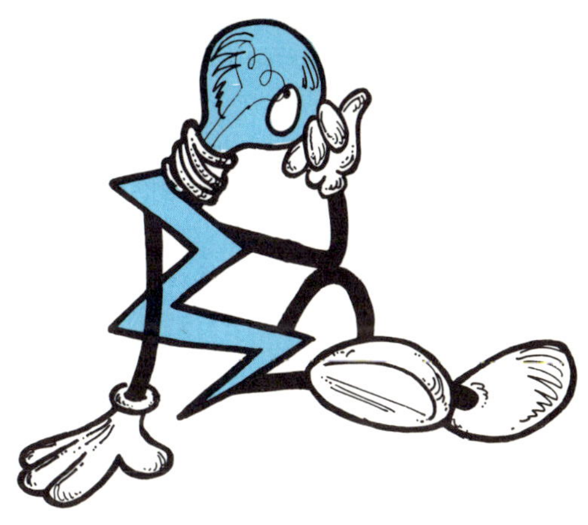

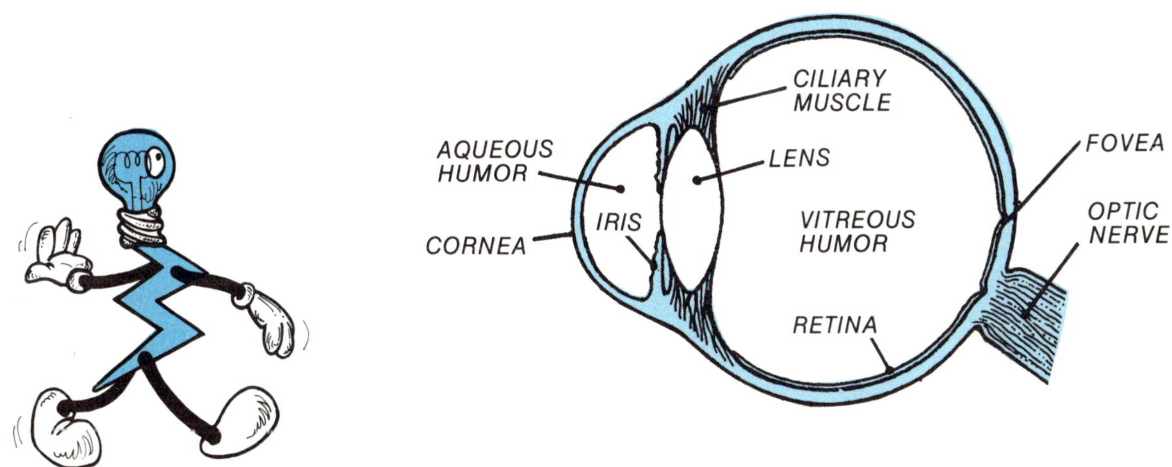

ANOTHER CHALLENGE TO YOUR THINKING

We have studied the eye. It is made of many parts. Each part is exactly the part needed. They are arranged in perfect order. If one part was out of place the eye would not work. How did all this happen in just the right order? What about the parts? Who or what made or developed them?

Do you have a transistor radio? Have you taken the back off to see how all the parts are connected together? Were these parts made by chance? Did the process of trial and error do it? Were they put together by a person? We know that the transistor radio was the work of people. Can this same reasoning be used for the eye.

visualize (vizh′ ü uh līz) to see a picture in your mind

transistor (tran zis′ tuhr) electronic device that transfers electrical signals

People who investigate things, plan experiments to test their ideas, and visualize ways to use the ideas to help solve problems are called "scientists." The scientist is able to understand things in the world by using facts and laws he has tested and found true. We have the transistor radio and many other useful products because of his work.

CREATION

The creationist says a design implies a designer. When he looks at the eye, he sees a design. This makes him think that someone with a mind carefully thought it through. Each part has its own design and is set in perfect order. Parts with design and set in perfect order have never been observed as a product of chance and long periods of time. The eye is a living thing. No one has ever observed a living thing coming from a nonliving thing. The evidence from observation and facts support the intelligent creation of the eye.

EVOLUTION

The evolutionist looks at the eye and says it has a design made by nature. He says science can't deal with a creator. The processes of nature are time, chance, and opportunity. The eye was made by nature through gradual changes over millions of years. The eye developed from a simple few-celled eye to the complicated eye of today. Once a design that works well has evolved, it is not likely to change. This supports the random evolution of the eye.

evidence (ev´uh duhns) facts that give proof

What do you think? Does the eye look like it was designed by a Creator or is it the result of time,[12] chance, and opportunity?

I hope the wonders of the eye have fanned your enthusiasm to think about the world and to investigate other parts of it. I have enjoyed shining my light ray through the eye for you and talking about this fantastic light receiving instrument. I hope you will remember what I have taught you. Have a happy day, and take care of your wonderful light receiver!

GLOSSARY

aqueous humor: fluid between the cornea and the lens of the eye

binocular: two pictures blend into one to see depth and shape

charges—plus, minus: alternating positive and negative charges in the nerve cells that propel the impulse to the brain

ciliary muscle: muscles controlling the lens

compound eye: eye with many facets and no retina

cones: nerve cells in the retina that pick up color

connectors: nerves that go from one part of the body to another

cornea: outside surface of the eye we see

creation: the idea that all life forms are varieties of kinds created by a master designer

crystalline: colorless, transparent material

decode: to give understanding

design: to plan with a purpose

evidence: facts that give proof

evolution: the idea that simple life forms can change to complex ones over long periods of time

focus: (a) point at which rays of light come together, or separate to make an image; (b) to look at something

fovea: area in retina where there are more cones and where vision is the clearest

gland: group of cells that do a special work

hypothesize: to take the evidence and make an explanation

image: just like something else

iris: colored part of the eye

lens: a clear, glass-like material that forms an image by focusing rays of light

lubricate: to make smooth or slippery

model: viewpoint

nerve cells: the cells in the body that transmit messages to the brain

optic nerve: a bundle of nerves that go from the back of the eye to the seeing center of the brain

pigment: color

pupil: hole in the iris of the eye

receiving: taking in light

receptor: a cell that takes in light

reflected: bounced back

retina: inner lining of the back of the eyeball

rods: nerve cells in the retina that pick up dim light

scientist: a person learned in science

simple eye: contains one lens that concentrates a light image upon a retina (on back of eye ball)

superpositioned compound eye: eye with many facets and a retina

transistor: electronic device that transfers electrical signals

transparent: material you can see through

visualize: to see a picture in your mind

vitreous humor: clear, jelly-like material in the eyeball to give it the required shape

REFERENCES

1. **McGovern, Ann** *The Human Body,* Random House, 1965, p. 31.
2. **McWilliams, John R.** *World Book Encyclopedia,* Vol. 6, p. 363, 1975, Field Enterprises Educational Corp., Chicago, Ill.
3. **Levi-Setti, Riccardo** *Fossils Magazine,* Feb. 2, 1974, p. 73.
4. **Oram, Raymond F.** *Biology Living Systems,* Merrill Pub. Co., Columbus, Ohio, 1973, p. 318-319.
5. **Binney, Ruth** Editor, *Body and Mind,* Rand Mc Nally, New York, 1976, p. 83.
6. **Ovam, Raymond F.** p. 318-319 Wallace, Robert A., p. 272-274.
7. **Sheeley, Phillip** *Cell Biology,* Wiley & Sons, New York, 1980, p. 182-183.
8. **Mader, Sylvia S.** *Inquiry Into Life,* William C. Brown Co., 1976, p. 343.
9. **McWilliams, John R.** *World Book,* Vol. 6, p. 362.
10. **Land, Dr. Michael** "Nature as an Optical Engineer," *News Science Magazine,* Oct. 4, 1979, p. 10.
11. **Land, Dr. Michael** p. 10.
12. **Bliss, Richard** *Fossils: Key to the Present,* CLP, San Diego, 1980, p. 68.

RESOURCE BOOKS

Binney, Ruth, *Body and Mind,* Rand McNally, Chicago, Ill., 1976.

Cosgrove, Margaret, *Wonders of Your Senses,* Dodd, New York, 1958.

Encyclopedia Brittanica, William Benton Publishers, Chicago, Illinois, 1974.

Fishbain, Morris, *Illustrated Medical and Health Encyclopedia,* Stuttman Co., New York, 1956.

Gilmour, Ann, *Understanding Your Senses,* Ward, London, England, 1963.

Heimler, Charles H., *Focus on Life Science,* Merrill, Columbus, Ohio, 1977.

Heimler, Charles H., *Principles of Science,* Merrill, Columbus, Ohio, 1975.

Jacobson, Willare J., *Investigating in Science,* American Book, New York, 1965.

Johnson, Dr. Gary, Ear, Nose, Throat, and Plastic Surgeon, tape, Mt. Vernon, Washington, 1979.

McWilliams, John, "Eye," *World Book Encyclopedia,* Field Enterprise, Chicago, 1969.

Moore, John N., *Biology: A Search for Order in Complexity,* Zondervan, Grand Rapids, Michigan, 1970.

Rothenburg, Robert E., *New Medical Encyclopedia for Home Use,* Abradale, New York, 1976.

APPENDIX

Laboratory Experiments

I. Make a vocabulary book that has the following:
   A. Make a cover. Draw or find a picture to decorate it. Use some ideas from the unit.
   B. Make a title page with a picture.
   C. Contents of the book.
      1. Choose twenty words from the glossary list.
      2. Define each word.
      3. Write them in syllables.
      4. Draw a picture of the idea of the word.

EXAMPLE

PUPIL (pu-pil) A pupil is the hole in the iris.

EXTRA CREDIT

Draw a picture to show light passing through the pupil to reach the retina.

II. EXPERIMENTS
   A. The pupil of the eye
      1. Look at your eye (or someone else's) in a bright light. Notice the black circle in the middle. Watch closely and you may see it swell or shrink just a little. Have your friend look at a pencil you hold near your nose. Watch the pupil in your friend's eye become smaller.
      2. Go into a room where the light is just bright enough for you to see. Look at your friend's eye again. What happened to the large circle? Turn on a light. Watch the circle.
         a. Record your observations.
         b. What conclusions can you make about your observations?
   B. Focus the eye

      Why can't you see close things and far away things at the same time?

EXPERIMENT:

Hold up your index fingers (pointing fingers). Move them out in front of you as far as possible about six inches apart. Look at all the detail on them. You can see knuckles, wrinkles, nails, scratches, lines, color of nails, and dirt under the nails . . .

Move your right finger toward your nose, but keep watching the left finger. When the right finger is almost touching your nose, how does it look?

OBSERVATION

The finger that you are *not* focusing on looks blurry. When you keep your eye focused on the finger you are moving toward you, the eye feels strained, as the lens of the eye thickens to keep the focus clear.

PRINCIPLE

You can only focus on one thing at a time. All other things are out of focus. The eye feels the strain as the lenses thicken to follow the moving finger to the nose.

C. Blind Spot

You can find the blind spot where the nerves enter the eye by following the experiment given below.

1. Draw a circle about ½" or 1 3/10 cm. in diameter.

2. About 1½ inches or 3.8 cm. to the left of the circle make a small cross (+). Each line of the cross is ¼ inch or .6 (6/10) cm.

3. Focus your left eye on the circle 12 inches or 30½ cm. away.

4. Look steadily with your left eye on the circle. You will be able to see the cross out of the corner of your eye.

5. Move the paper slowly toward you. Keep your left eye on the circle. There will come a point when the cross disappears. At this point the image of the cross is falling on the "blind spot" of your left eye. It is falling on the part of the retina where the optic nerve is leaving the eyeball to travel to the brain.

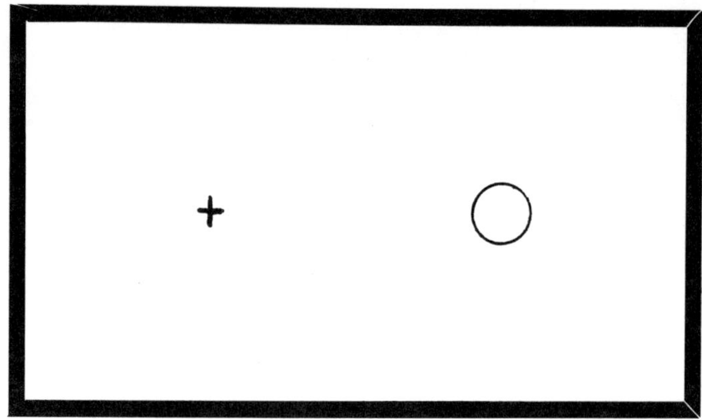

D. Two Images

Close one eye and look at your finger held out in front of you. Close the other eye and observe the same object. Do you see the same background? Also observe a distant object. Keep the pen held out in front of you. How many pens do you see? This demonstrates the two images that focus on the retina from each eye.

E. Seeing depth or third dimension

Both of our eyes are in front of the head. A horse or rabbit has one on each side of its head. One eye has an entirely different picture than the other eye. The two fields of vision barely overlap.

EXPERIMENT

1. Hold up a thick book about 25.2 cm (10") from your eyes. Look at the binding with both eyes open.
2. Now look at it with the left eye alone.
3. Now look at it with the right eye alone.
4. The two views are slightly different, but in the brain the images become one image to give depth and shape.
5. Close one eye and try to thread a needle.
6. Hold a pen that has a cap out in front of you as far as possible and close one eye. Try to replace the cap. It is easier to replace the cap using two eyes.